FASCINATING SCIENCE PROJECTS

LIGHT

Sally Hewitt

COPPER BEECH BOOKS
Brookfield • Connecticut

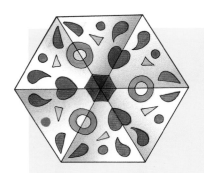

© Aladdin Books Ltd 2002
Produced by
Aladdin Books Ltd
28 Percy Street
London W1T 2BZ

ISBN 0–7613–2818–1 (lib. bdg.)
ISBN 0–7613–1736–8 (pbk.)

First published in the United States
in 2002 by
Copper Beech Books,
an imprint of
The Millbrook Press
2 Old New Milford Road
Brookfield, Connecticut 06804

Designers:
Flick, Book Design & Graphics
Pete Bennett

Editor:
Harriet Brown

Illustrators:
Ian Thompson,
Catherine Ward and Peter Wilks—SGA
Cartoons: Tony Kenyon—BL Kearley

Consultant:
Dr. Bryson Gore

Cataloging-in-Publication data is on
file at the Library of Congress

10 9 8 7 6 5 4 3 2 1

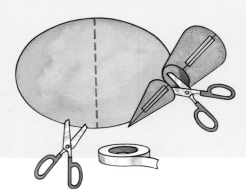

Contents

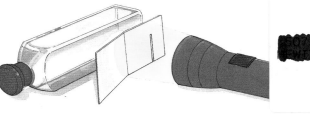

What is light? 6
Make your own light and see how sunlight and moonlight are different

Shadows 10
Learn how to put on a puppet show using only your hands

Reflection 14
Learn to write in code and find out how to make a kaleidoscope

Refraction 18
Discover what makes light change direction

Lenses 22
Magnify the moon and see how to make your own microscope

Colors of light 26
Split light to make the colors of the rainbow and create a sunset

Mixing light 30
See how colored light can be mixed to make white light

Colors and dyes 34
Dye cloth using beets and draw a picture using only dots

Taking pictures 38
Make your own camera to learn how X rays work

Weird light 42
Find out how fiber-optic cables work and see how sugar can create light

Glossary 46

Index 48

Introduction

In this book, the science of light is explained through a series of fascinating projects and experiments. Each chapter deals with a different topic on light, such as shadows or colors, and contains a major project that is fully supported by simple experiments, "Magic panels," and "Fascinating fact" boxes. At the end of every chapter is an explanation of what has happened and what this means. Projects requiring sharp tools or the use of heat should be done with adult supervision.

This states the purpose of the project

METHOD NOTES
Helpful hints on things to remember when carrying out your project.

Materials
In this box is a full list of the items needed to carry out each main project.

Figure 2

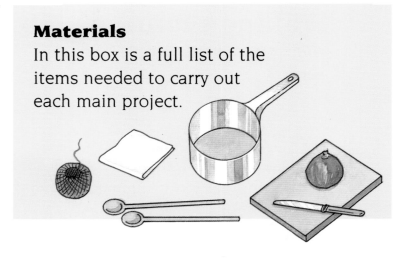

1. The steps that describe how to carry out each project are listed clearly as numbered points.
2. Where there are illustrations to help you understand the instructions, the text refers to them as Figure 1, etc.

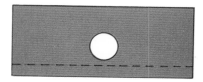
Figure 1

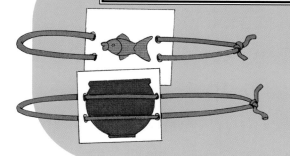

These boxes contain an activity or experiment that has a particularly dramatic or surprising result!

WHY IT WORKS
You can find out exactly what happened here, too.

WHY IT WORKS

These boxes, which are headed either "What this shows" or "Why it works," contain an explanation of what happened during your project, why it happened, and the meaning of the result.

Fascinating facts!
An amusing or surprising fact related to the theme of the chapter.

Where the project involves using a sharp knife, heat, or anything else that requires adult supervision, you will see this warning symbol.

The text in these circles links the theme of the topic from one page to the next in the chapter.

What is light?

Light is a kind of energy. We cannot see without it. Light from the Sun lights up Earth by day. At night, when it is dark, we turn on electric lights or light candles to see by. The Sun, electric light bulbs, and candles are all luminous, which means they give out light of their own. Light travels very fast in straight lines called rays. When rays of light hit something solid like you, a dark patch called a shadow is formed where the light cannot reach.

Explore how a light bulb glows

METHOD NOTES
Use tweezers to pull out a single wire from soapless steel wool.

Materials
- two 1.5 volt batteries
- 2 pieces of insulated wire
- modeling clay
- 2 crocodile clips
- a single wire from steel wool
- a tray of sand

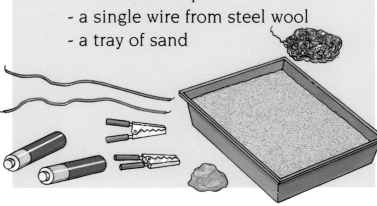

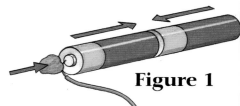

Figure 1

1. Place the two batteries together with a positive (+) terminal touching a negative (-) terminal (Figure 1).
2. Attach the exposed end of one piece of insulated wire to one of the battery terminals using a piece of modeling clay (Figure 1).

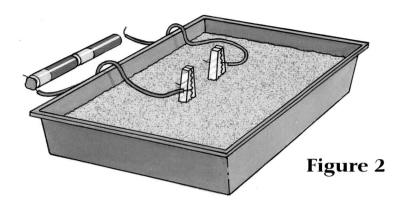

Figure 2

3. Lay the end of the other insulated wire near the other battery terminal.

4. Stand the crocodile clips in the sand tray and attach the other ends of the insulated wires to the clips (Figure 2).

5. Attach the single wire from the steel wool between the two crocodile clips (Figure 3).

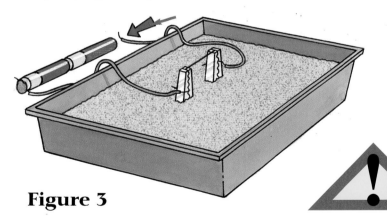

Figure 3

6. Complete the circuit by touching the loose, exposed end of the wire to the battery terminal. Watch the steel wool glow, flare up, and break as electricity flows through it (Figure 4). Stand back as the wire flares and don't touch it.

WHAT THIS SHOWS

The steel-wool wire makes up part of an electric circuit. It is so thin that it becomes red hot and glows when electricity passes through it.

The wire catches fire and burns out in air. The fine wire in an electric light bulb, called a filament, is made of a metal called tungsten. A light bulb is filled with a gas called argon that lets the filament glow brightly for a long time without burning out.

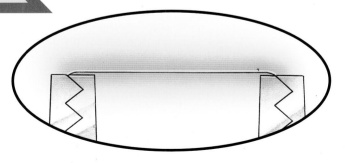

Figure 4

Some luminous things give out brighter light than others. The brightness of light is called its intensity. Light from a flashlight is more intense than candle light. As light spreads out, it becomes less intense.

What is light?

SUNLIGHT AND MOONLIGHT

In a dark room, shine the flashlight onto a postcard placed inside a box (Figure 1). How well can you see the picture? Make a cone from half a circle of cardboard (Figure 2) and attach it to the end of the flashlight to direct the beam. Now, crumple a piece of white paper into a ball for the moon. Stick a length of string to the ball with tape (Figure 3). Hang the ball in front of the box and shine the beam onto the ball to make it glow (Figure 4). How well does the ball light up the picture?

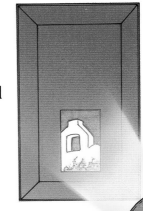

Figure 1

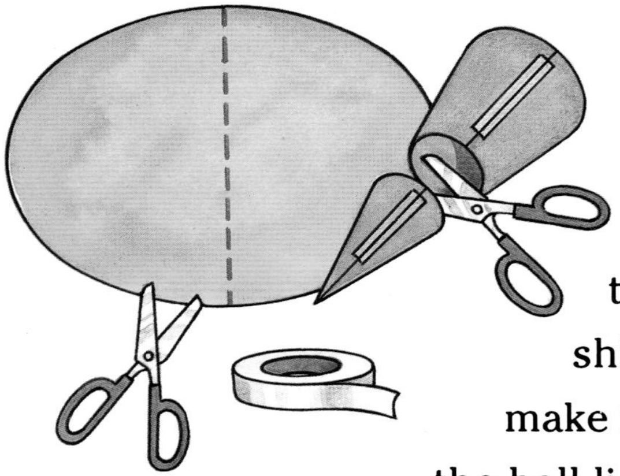

Figure 2

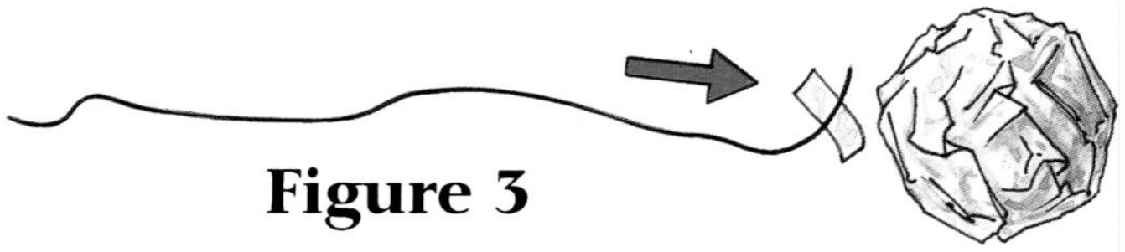

Figure 3

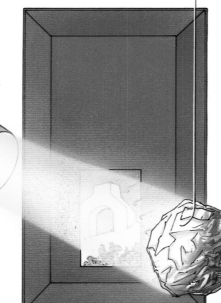

Figure 4

WHAT THIS SHOWS

Sunlight is brighter than moonlight. The moon has no light of its own. Moonlight is reflected light from the Sun. Light from the flashlight is reflected off the ball of paper. This reflected light is weaker than light from the flashlight.

LIGHT AND SHADOW

In a dark room, carefully light a candle and place objects all around it. See how each object around the candle casts a shadow in a different direction.

This happens because a candle spreads light all around it. The beam of a flashlight shines in the direction where you point it. If you shine a flashlight at the objects, all the shadows will point away from it.

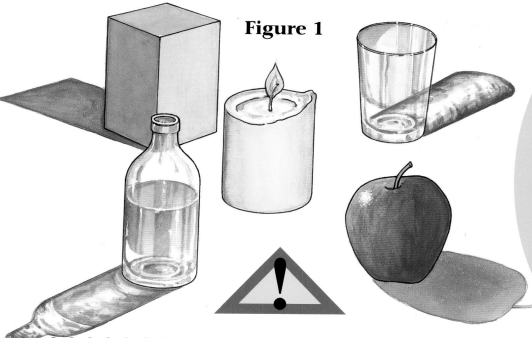

Figure 1

The Sun is the Earth's greatest source of light. We need light to see, so we use luminous objects such as electric lights and candles to light up the darkness.

Shadows

When a solid object lies in the path of a ray of light, it creates a dark patch called a shadow where the light cannot reach. Night is created by the Earth's shadow. As the Earth spins around in space, half of it is always facing the Sun and the other half is in shadow. When the part of Earth you are on is facing the Sun, it is day; when it is facing away from the Sun, it is night.

Make a sundial to see how shadows work

METHOD NOTES
Use a compass to place the sundial so that the dowel is on the south edge of the cardboard.

Materials
- modeling clay
- a rectangular piece of thick cardboard
- glue
- a length of dowel
- waterproof paints
- a ruler
- a skewer or knitting needle
- a compass

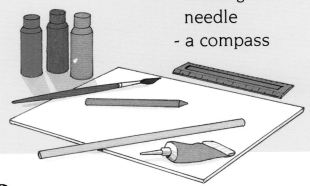

1. Put a piece of modeling clay under the middle of one of the long edges of cardboard. Push a skewer or knitting needle through the cardboard and into the clay to make a hole.

2. Glue the dowel into the hole so it stands upright, then remove the clay (Figure 1).

3. Use waterproof paint to decorate the sundial (Figure 2).

Figure 1

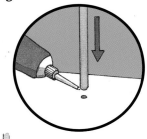

Figure 2

4. On a bright sunny morning put the sundial outside. On the hour exactly, draw a line along the shadow made by the dowel using a ruler, and write the time next to it (Figure 3).

5. Repeat this every hour on the hour throughout the day. You will end up with a series of lines on the sundial at even intervals (Figure 4).

6. Use your sundial to tell the time. It will work only on sunny days, and you must always place it in exactly the same position.

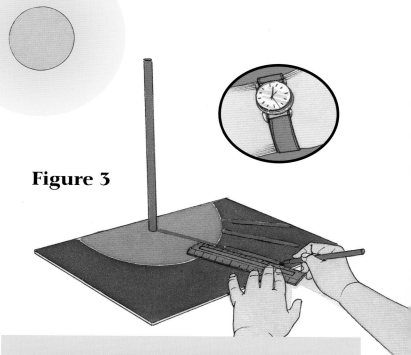

Figure 3

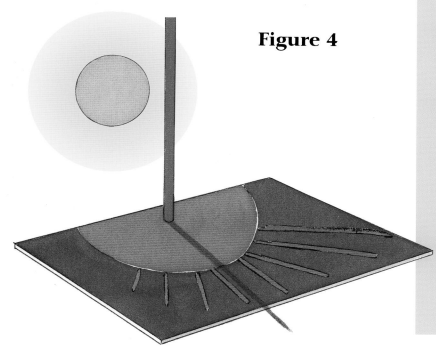

Figure 4

WHAT THIS SHOWS

The dowel casts a shadow where it blocks light from the Sun. The position of the shadow on the sundial changes as the Sun moves across the sky.

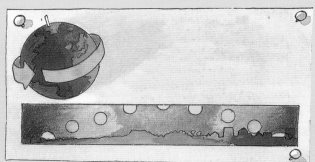

In fact, the Sun stays still in space and the Earth revolves around it. To us, it appears that the Sun rises in the east. It reaches its highest point at midday. Then it sets in the west.

Shadows

A small light creates a very dark shadow called an umbra. When a large source of light lets some light through, the edge of the shadow looks gray and is called a penumbra.

HOW SHADOWS CHANGE THROUGHOUT THE DAY

The position of the Sun in the sky affects the length and position of your shadow. On a sunny morning, stand outside on a hard surface.

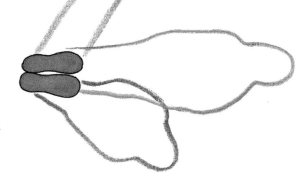

Get a friend to draw around your shadow with chalk. Stand in the same place at midday when the Sun is nearly overhead and repeat the process. Your shadow is longer in the morning and afternoon when the Sun is low in the sky. It is shorter at midday when the Sun is overhead.

THE AMAZING PUPPET SHOW
Make a puppet show of birds and dogs only using your hands

In a dark room, shine a powerful flashlight onto a wall. Copy these shapes with your hands. Put your hands in front of the light and move them to flap wings and wiggle ears.

WHY IT WORKS
Light cannot shine through your hands, so a shadow of the shape, or silhouette, of your hands is cast onto the wall. The nearer your hands are to the light, the bigger the shadows will be.

Figure 1

OPAQUE, TRANSLUCENT, AND TRANSPARENT

Find a piece of clear Plexiglass, some tissue paper, and a piece of cardboard. Shine a bright flashlight onto each one in turn. No light will pass through the cardboard and you will see a shadow (Figure 1). Only a little light will pass through the tissue (Figure 2), whereas it will shine right through the clear Plexiglass (Figure 3).

Figure 2

Figure 3

WHY IT WORKS
The clear plastic is transparent, which means light can shine through it. The tissue paper is translucent, which means it lets a little light through. The cardboard is opaque, and lets no light through. Clouds are translucent and only let a little sunlight through, so shadows are faint on a cloudy day.

Light cannot shine through things that are opaque, so opaque things cast a shadow. Using a sundial, we can tell the time on a sunny day by the changing shadows.

Reflection

As light travels in straight lines, it strikes objects in its path. It bounces off the objects rather like a ball bouncing off a wall. We see things because light bounces off them. We call this reflection. Rough surfaces scatter light that is reflected in all directions. Smooth, flat surfaces reflect light in one direction. White things show up because white reflects light. Dark colors are hard to see because they absorb, or take in, light.

Make a kaleidoscope to explore reflected light

METHOD NOTES
If you use mirror board, ask an adult to help you cut it with a craft knife.

Materials
- 3 rectangles of mirror board or 3 mirrors 5 1/2 x 2 1/2 in
- cardboard
- tracing paper
- colored paper or sequins
- transparent tape
- a sharp pencil
- glue

Figure 1

1. Tape the 3 rectangles of mirror board, or the mirrors, together with the mirrors facing inward to make a triangular prism (Figure 1).

2. On the cardboard, draw around one end of the prism to make a triangle (Figure 2). Cut it out and make a hole in its center with the tip of a sharp pencil.

3. On the tracing paper, draw two triangles in the same way. Draw flaps about 1/4 in wide along each edge (Figure 3) and cut the triangles out.

4. Glue the flaps of the two triangles together along two sides only to make a little pocket (Figure 4).

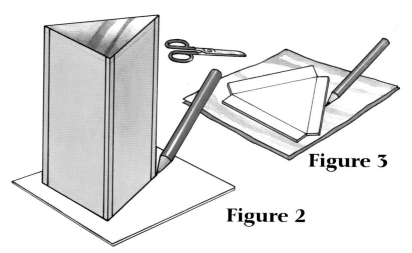

Figure 3

Figure 2

Figure 4

Figure 5

5. Fill the pocket with sequins or bits of colored cardboard (Figure 5).

6. Tape the pocket to one end of the prism (Figure 6) and the cardboard triangle to the other. Hold the kaleidoscope up to the light and look through the hole. Turn it to change the patterns you see (Figure 7).

WHAT THIS SHOWS

Mirrors are usually made of smooth glass with shiny metal on the back. We see a reflection when light from an object bounces off the mirror into our eyes. Light from the sequins bounces between the mirrors, and we see their reflections over and over again.

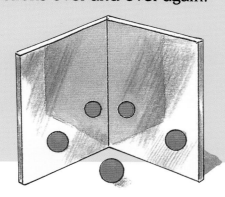

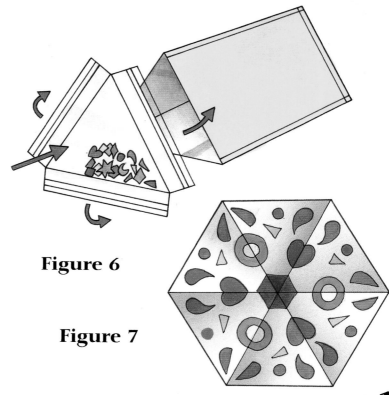

Figure 6

Figure 7

Reflections

Days are bright because sunlight is reflected and scattered in all directions by the Earth's atmosphere.

HOW LIGHT BOUNCES

Cut out a circle of aluminum foil large enough to cover the beam of a flashlight. Make a hole in it with a pencil (Figure 1). Cover the flashlight beam with foil (Figure 2) to create a narrow, bright beam of light. In a dark room, shine the flashlight onto a mirror. You will see a beam of light reflected off the mirror. Ask a friend to hold up a ball. Take aim and try to hit the ball with the beam of reflected light (Figure 3). You change the angle of the beam by moving either the mirror or the flashlight (Figure 4).

Figure 1

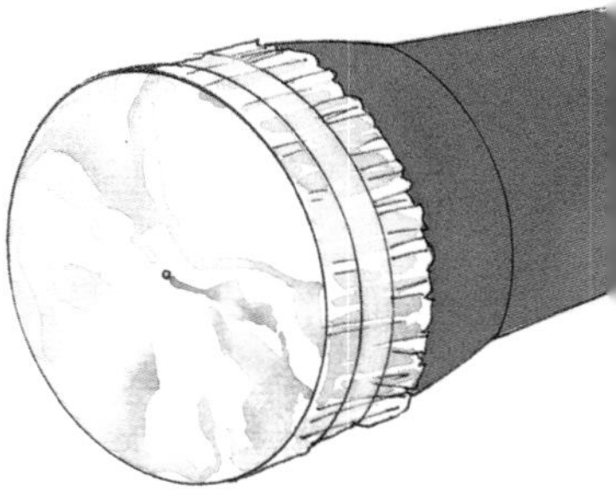

Figure 2

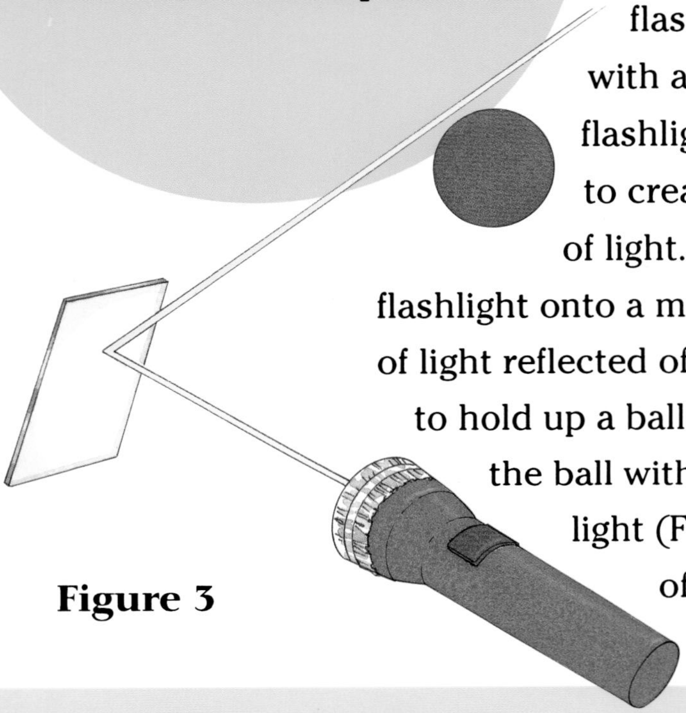

Figure 3

WHY IT WORKS

If you shine a beam of light straight at a smooth surface like a mirror, it will bounce straight back. If the light hits the surface at an angle, it will be reflected back at exactly the same angle. By looking at where the ball is in relation to you and the flashlight, you can calculate the angle you need to make the beam of light reflect onto the ball.

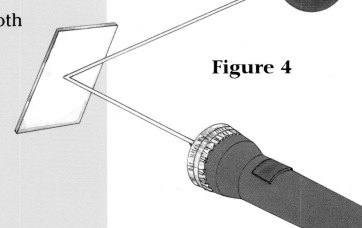

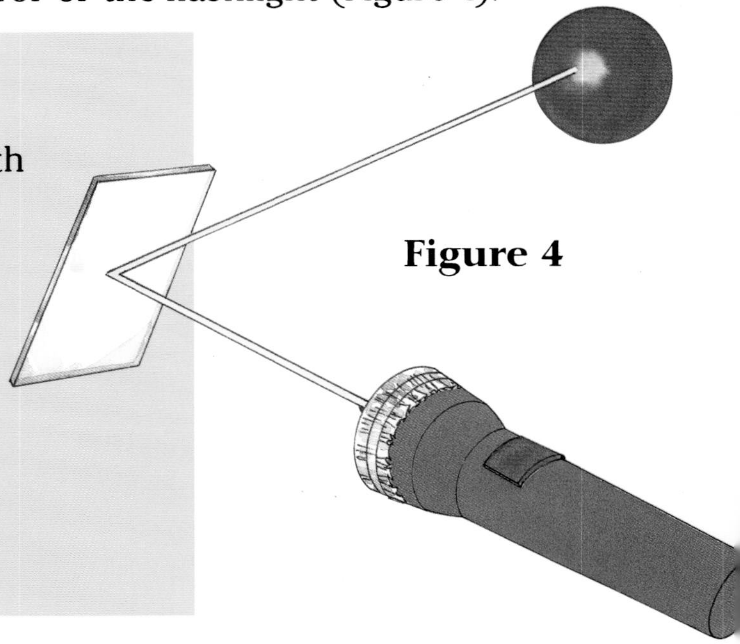

Figure 4

THE AMAZING FIERY FINGER
See how to fool your friends with reflections

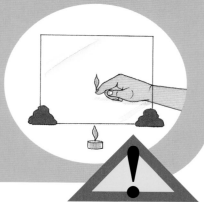

Set a sheet of Plexiglass upright in modeling clay. Put a small candle on one side of the plastic and light it. Angle your finger on the other side of the Plexiglass so that it looks like the flame is coming from your finger.

HOW IT WORKS
Your friend sees the reflection of the candle flame on the shiny plastic, creating the illusion that your finger is on fire.

MIRROR WRITING

Write your name on a piece of paper and put it in front of a small mirror. Copy what you see in the mirror. Now, put this writing in front of the mirror, and you will see a reflection of your name. Mirror writing looks like secret code, so you can use it for secret messages and use a mirror to decode it. Reflections in the mirror look the wrong way around because they bounce straight back off the mirror, and your image is reversed. Your left hand looks like your right hand in the mirror.

Light is reflected in different ways. It bounces straight off smooth surfaces, and it scatters when it hits rough surfaces. We see objects when this reflected light goes into our eyes.

Refraction

Light going through empty space with nothing in its way will travel in a straight line. But when light goes from air through something transparent, such as glass or water, it is refracted—that means that its direction changes. This happens because light travels at different speeds through different materials. When light passes through raindrops or a triangular-shaped piece of glass called a prism, it is refracted and split into colors. Light is always on the move; it never stays still.

See light refract as it passes through water

METHOD NOTES
Use a jar or container with straight sides.

Materials
- a piece of cardboard
- scissors
- a glass jar with a screw top
- a few drops of milk
- a flashlight
- water

Figure 1

1. Cut a narrow slit in one end of the cardboard (Figure 1) and bend it to stand up on its long side.

2. Fill the glass jar with water and mix in the milk (Figure 2). Screw the lid on tightly.

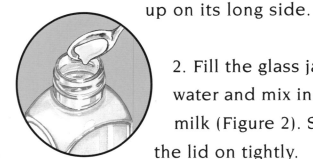

Figure 2

Figure 3

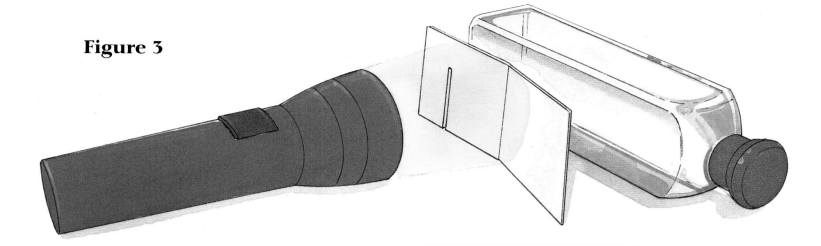

3. In a dark room, lay the jar on its side on a table.

4. Stand the cardboard on the table next to the glass jar. Then shine the flashlight through the slit into the jar (Figure 3).

5. See the beam of light bend as it goes through the water, and then bend the other way as it passes back into the air.

Half fill a clear plastic bag with water and hold it up to a window. What can you see through the bag? Can you see rainbow colors? Try to work out what happens to the light as it passes through the water.

WHAT THIS SHOWS

Light travels faster through air than through glass or water. As it goes from the air into the water, it slows down slightly and bends.

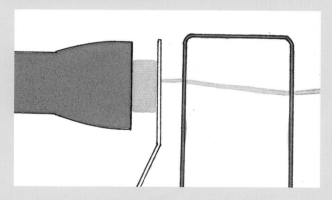

This is called refraction. The beam of light bends as it enters the water. It travels in a straight line through the water, and then bends again as it leaves the water and goes back into the air.

Refraction

SEE A BEAM OF LIGHT

Spread some newspaper on the floor in a dark room. Gently sprinkle flour through a sieve onto the paper. Get a friend to shine a flashlight through the flour as you are sifting the grains, and you will be able to see the path of the beam of light (Figure 1).

Figure 1

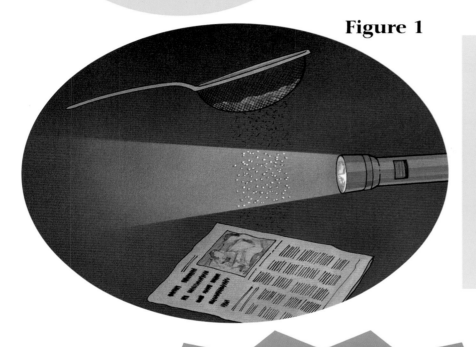

WHAT THIS SHOWS

When sunlight shines through dust or mist, when headlights shine through rain, and when your flashlight shines through grains of flour, the drops of water and grains are lit up, letting you see the straight path of the light rays.

Disappearing visions

Thirsty desert travelers often think they see a pool of water (an "oasis") that disappears as they get near. What they really see is a mirage, a refraction of blue sky shimmering in the hot air near the ground.

THE AMAZING DISAPPEARING COIN
See how refraction can make things appear

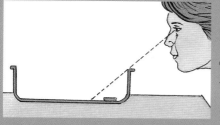

Put an empty bowl on a table and put a coin in the bottom. Secure it in place with modeling clay. Look at the coin and move backward until it just disappears from view. Stay in the same position and get a friend to pour water into the bowl. Watch the coin reappear, as if by magic!

WHY IT WORKS
The water bends the light from the coin and brings it back into view.

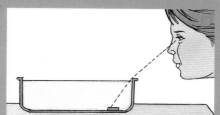

BENDING PENCILS

Half fill a glass with water and stand a pencil in it. When you look at the pencil through the side of the glass, it seems to bend where it goes into the water. Light rays from the pencil bend as they leave the water, making it look bent, although when you pull it out, you will see it is really still straight.

Figure 1

Light never stops moving in straight lines, called rays. It travels at different speeds through air, water, and glass. As it goes from one of these materials to another, it changes direction, or refracts.

Lenses

We cannot change what we see with our eyes, but we can use lenses in glasses, cameras, microscopes, binoculars, and telescopes to make things look clearer, bigger, or smaller. A concave lens curves inward toward the middle and makes things look smaller. A convex lens bulges outward toward the middle and makes things look bigger. In binoculars, convex lenses magnify what you see, making distant objects appear to be much closer than they are.

Make a microscope with a drop of water for a lens

METHOD NOTES
Ask an adult to cut the bottle with a craft knife for you.

Materials
- a transparent plastic bottle
- a craft knife
- a small mirror
- modeling clay
- a drinking straw
- a drop of water
- scissors
- a hair

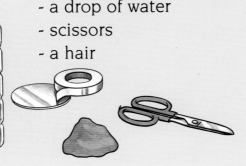

1. Cut the top off the bottle. Cut narrow strips from two opposite sides of the bottle and keep them (Figure 1).
2. Cut two horizontal slits in the other two opposite sides, near the top. Push each end of one of the strips into the slits to make a platform (Figure 2).

Figure 1

22

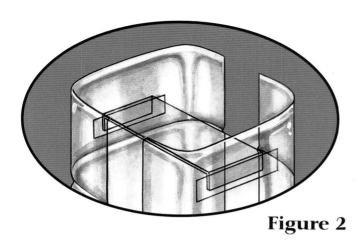

Figure 2

3. Prop a mirror at an angle on the modeling clay in the bottom of the bottle to reflect light upward (Figure 3).
4. Dip a drinking straw in water and block the top hole with your finger. Take your finger off the straw to release one drop of water onto the platform.
5. Put the hair on the other strip and hold it under the drop of water. Look at it through the drop of water and see that it looks bigger (Figure 4).

Figure 3

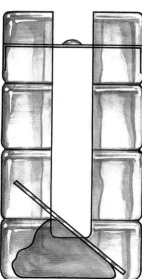

Look at other things— a tiny piece of newspaper with small print, a grain of sugar, and a petal, for example.

WHAT THIS SHOWS

The drop of water acts like a tiny convex lens (a). The mirror reflects light up onto the hair.

(a)   (b)

Light rays from the hair bend and converge—come closer together—as they pass through the drop of water, and they make the hair look bigger. Light rays bend and spread out as they pass through a concave lens (b), making things look smaller.

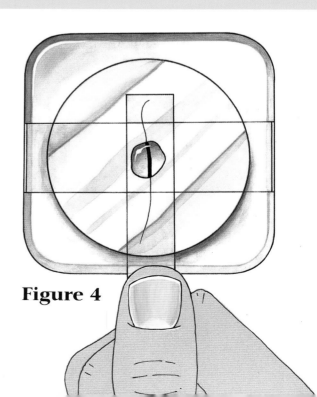

Figure 4

Lenses

Concave lenses in glasses help nearsighted people, and convex lenses help farsighted people, to see clearly. Contact lenses work in the same way, but are put straight into the eye.

CONCAVE AND CONVEX LENSES

Borrow a pair of glasses from someone who is nearsighted to use as a concave lens. Use a magnifying glass as a convex lens. Cut a hole in a piece of cardboard and bend the cardboard so it stands up (Figure 1). Tape a comb over the hole (Figure 2). Put the magnifying glass in front of the hole (Figure 3). In a dark room, shine a flashlight through the hole so that the rays shine on a book. See the rays created by the comb focus to a point. Replace the magnifying glass with the glasses, and see the rays spread out (Figure 4).

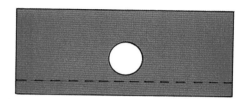

Figure 1

Figure 2

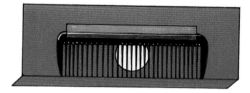

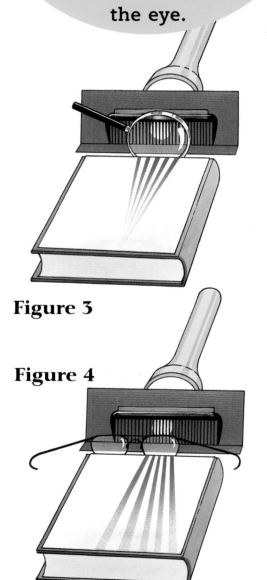

Figure 3

Figure 4

WHY IT WORKS

A convex lens bends the rays so that they focus, or come together, at a point, making things look larger. A concave lens bends the rays so that instead of focusing, they spread out and make things look smaller.

Hold a pen a few inches from a shiny spoon so you can see its reflection in the side of the spoon that bends inward. In the reflection, you will see the pen upside down.

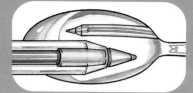

Slowly, move the pen nearer to the spoon so its reflection looks even bigger than the pen. Watch the reflection of the pen suddenly turn the right way up.

MAGNIFY THE MOON

A concave mirror, like a convex lens, magnifies objects. Put a concave makeup mirror in a window facing the moon. Hold a flat mirror so that you can see the reflection of the moon in the makeup mirror (Figure 1). Now look at that reflection through a magnifying glass to see a clear image. Convex and concave mirrors are used in telescopes to look at the stars (Figure 2).

Figure 1

Figure 2

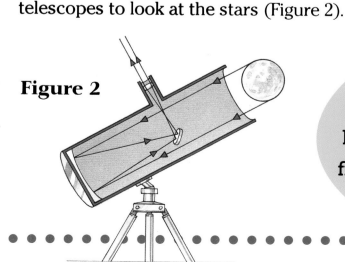

Concave and convex mirrors and lenses are used all around us every day to bend light and change the way we see things—from tiny creatures under a microscope, to distant stars through a telescope.

Colors of light

Light that we see from the Sun looks white, so we call it white light. In fact, light is made up of hundreds of colors called the spectrum. People call the colors of the rainbow red, orange, yellow, green, blue, indigo, and violet. White light splits into these seven colors when it passes through a triangular piece of glass called a prism. Sunlight splits when it passes through raindrops that act as tiny prisms, and we see a rainbow.

Make a rainbow to see the seven colors of light

METHOD NOTES
You can use a flashlight or a beam of sunlight for this experiment.

Materials
- black cardboard
- scissors
- a bowl of water
- a mirror
- modeling clay
- a smaller piece of white cardboard
- a flashlight

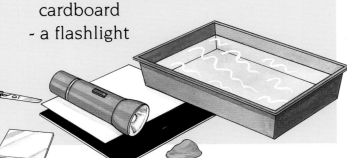

1. Cut a horizontal split just below the top edge of the black cardboard (Figure 1).
2. Bend the bottom of the cardboard so it can stand upright.
3. Half fill a glass bowl with water.

Figure 1

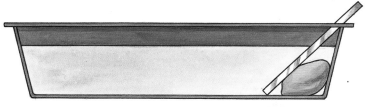

Figure 2

4. Angle a mirror so it is half in and half out of the water (Figure 2). Use modeling clay to keep it in place.

5. Stand the black cardboard with the white cardboard in front of it, with the slit facing the mirror (Figure 3).

6. Shine the flashlight through the slit at the mirror (Figure 3). Adjust the mirror until you see a rainbow on the white cardboard.

Figure 3

WHAT THIS SHOWS

A prism is created in the triangular shape between the mirror and the water. As the ray of light passes through the prism, each color travels at a slightly different speed and bends at a different angle. The white light splits into a spectrum, and you see a rainbow reflected onto the white cardboard.

What shape are rainbows?
If you stood on a mountain and looked down on a rainbow, you would see that it is actually a whole circle. It usually looks like an arch because we can only see part of it from the ground.

Colors of light

MAKE RAINBOW COLORS

Put a bowl of water in some bright sunshine, and sprinkle in a few drops of oil. Stir the oil gently, and see rainbow colors floating on the water.

Buy some bubble mixture, or make your own by adding bubble bath to water. Blow bubbles outside on a sunny day and see rainbow colors on their skin (Figure 1). Light is reflected between the thin layers of the bubbles' soapy skin and between the thin layers of oil on the water. The light is split and a spectrum is created in the soapy and oily skins.

Figure 1

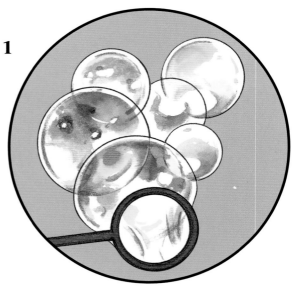

wavelength

DIFFERENT COLORS

Each color of light travels in tiny waves—all with a different wavelength. Waves are measured from the top of one wave to the next. Red has the longest wavelength, and violet has the shortest.

WHY THE SKY IS RED AT SUNSET

Stir a teaspoon of milk into a jar of water (Figure 1). Shine a flashlight sideways through the jar. The milky water looks blue (Figure 1). Move the flashlight so its light is shining through the jar toward you, and the water looks yellow (Figure 2). Stir in a little more milk and shine the flashlight—the water looks pink (Figure 3).

Figure 2

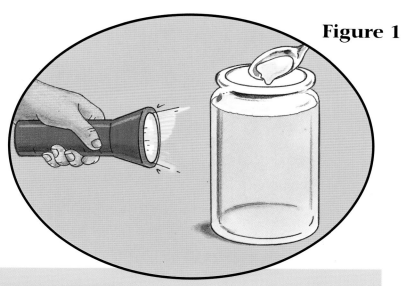

Figure 1

Figure 3

WHAT THIS SHOWS

The milky water looks blue as blue light is scattered out of the jar. Yellow light shone toward you looks yellow as none of it is scattered. In the third case, only red light can get through the cloudier liquid. At sunset, the low Sun travels through more particles in the atmosphere. Only red light gets through.

Sunlight splits into many colors. Tiny particles in the Earth's atmosphere scatter blue light, making the sky look blue in the day.

Mixing light

Light and color go together; you can't have one without the other. When light enters our eyes, we see color with light-sensitive cells inside our eyes called cones. These detect the three primary colors of light: red, green, and blue. The things around us appear to be different colors because of the color of light they reflect. A green leaf, for example, reflects green light into our eyes, and absorbs or takes in all the other colors.

Mix the primary colors of light

METHOD NOTES
This experiment works best if you do it with two friends.

Materials
- red, blue, and green cellophane
- scissors
- 3 cardboard tubes
- 3 flashlights
- transparent tape
- white cardboard

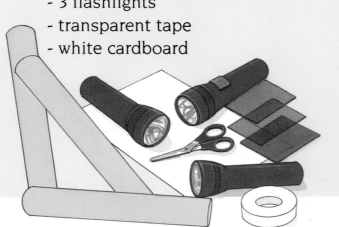

1. Cut out a circle of each color of cellophane to make them into three filters. The circles should be bigger than the end of the tubes (Figure 1).

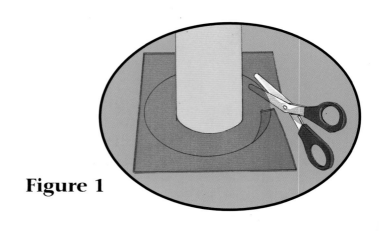

Figure 1

30

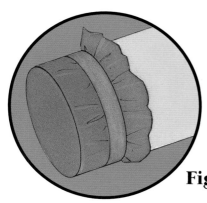

Figure 2

2. Tape a colored filter to the end of each tube (Figure 2).

3. Put the white cardboard on the floor. You and two friends need a flashlight and a tube each.

4. In a dark room, shine the flashlights down each of the tubes to make three pools of colored light on the white cardboard (Figure 3).

5. Move the lights around so they overlap and make new colors (Figure 3). Can you see a pool of white light where the three colors overlap?

Figure 3

Why not try making a color wheel? Cut out a circle of cardboard. Paint the seven colors of the rainbow in order on the cardboard. Push a pencil through the middle and spin the wheel. The colors mix and make white light.

WHAT THIS SHOWS

You have seen that light can be split into many colors. When all the colors are mixed together again, they make white light. Mixing two colored lights together makes different colors. Green and blue light make the pale blue light called cyan. Blue and red light make the pink light called magenta, and green and red light make yellow light. Colored filters on spotlights are used to create colored lighting effects on the stage.

Mixing light

When an object absorbs all the colors of light, it looks black. Objects that reflect all the colors of light into our eyes look white.

LOOK AT OBJECTS IN COLORED LIGHT

Cut a big hole in the lid of a shoebox (Figure 1). Cut out a small hole in one of the short sides of the box (Figure 2). Put different colored things, such as a banana, a tomato, and a green apple, in the box. Lay a sheet of red cellophane over the lid.

Figure 1

Figure 2

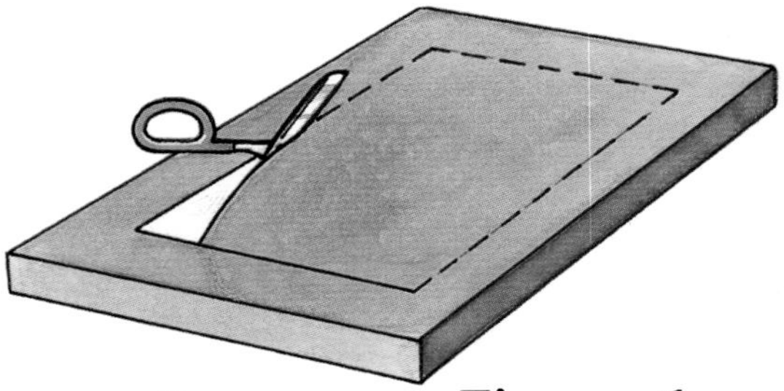

Shine a flashlight through the cellophane filter. Look through the hole in the side of the box and see how the red filter changes the colors of the fruit in the box (Figure 3). Red objects in the box look pale, and green ones look dark. Try using different colored filters.

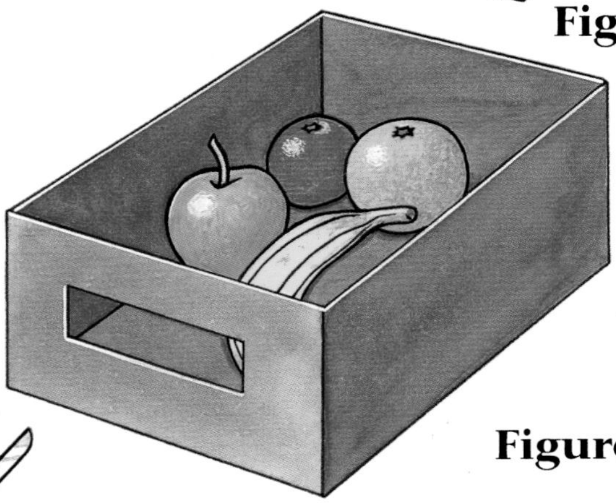

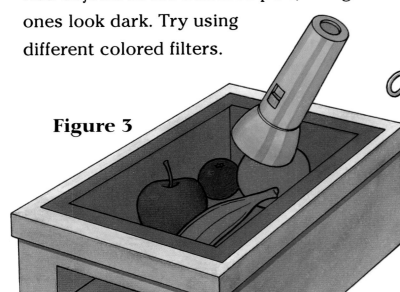

Figure 3

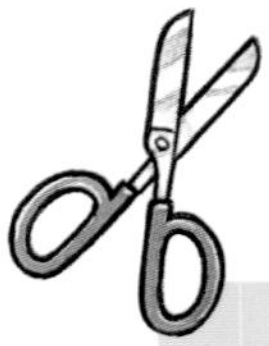

WHAT THIS SHOWS

A red filter allows only red light through, just as a green filter lets only green light through, and so on. Green objects look dark through the red filter because they absorb red light. They can't reflect green light because no green light comes through the filter.

A blue banana!
Some animals' eyes are able to see more colors than ours. Very often, these animals are brightly colored themselves. They use these colors for display and defense.

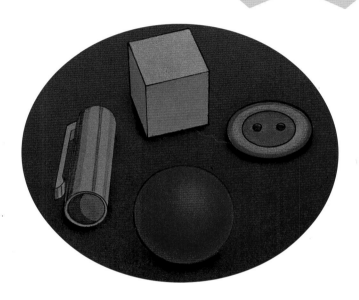

SEEING IN THE DARK

Look at a collection of colored objects in a darkened room, with just enough light to see by. Can you tell what colors they are? There is not enough light for your eyes to detect different colors. We see in dim light with cells in our eyes called rods, which detect only black and white.

TWILIGHT

Look out of the window at twilight—when the Sun has just sunk below the horizon. See how colors begin to fade as the light disappears.

White light can be split into the colors of the spectrum. Colored light can be mixed to make white light. Human and animal eyes are adapted to see color and light in different ways.

Colors and dyes

Plants, soil, rocks, and animals have natural colors called pigments. These materials can be ground into colored powders, and used to make paints and dyes. Natural and chemical pigments used in paints and dyes can make an enormous range of different shades of every color. The pigments reflect some of the colors of light into our eyes, and we see those colors.

Tie-dye cloth in a natural pigment

Materials
- a chopping board
- a knife
- beets
- a pan of water
- tongs or long handled spoon
- white cotton fabric
- string
- scissors

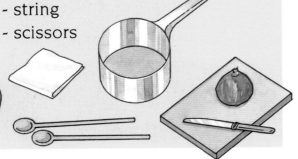

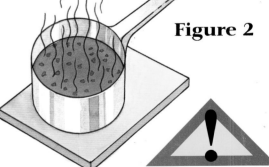

METHOD NOTES
You must wear an apron and rubber gloves for this project.

Figure 2

Figure 1

1. On a chopping board, carefully cut a beet into small cubes (Figure 1).

2. Put the cubes into a pan of water, bring it to the boil, and simmer for about 15 minutes until the water turns purple (Figure 2).

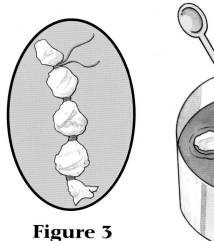

Figure 3

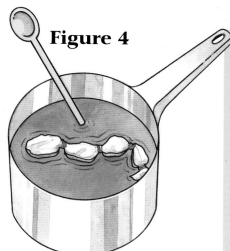

Figure 4

3. Take the pan off the heat and allow to cool.

4. Roll up the fabric and tie the string around it tightly in three or four different places (Figure 3).

5. Drop the fabric in the purple water and leave it in for a few minutes (Figure 4).

6. Take it out with the tongs (Figure 5), cut the string (Figure 6), and then hang it up to dry (Figure 7).

WHY IT WORKS

Vegetables contain colored pigments that go into the boiling water. Beets, red cabbage, and cherries contain red pigment. Spinach contains green pigment, and onion skins contain yellow pigment. Soaking the material in the colored water dyes it. The dye cannot reach the parts of the cloth you tie tightly with string, so they stay white.

Experiment with tying the string to make different patterns on the cloth.

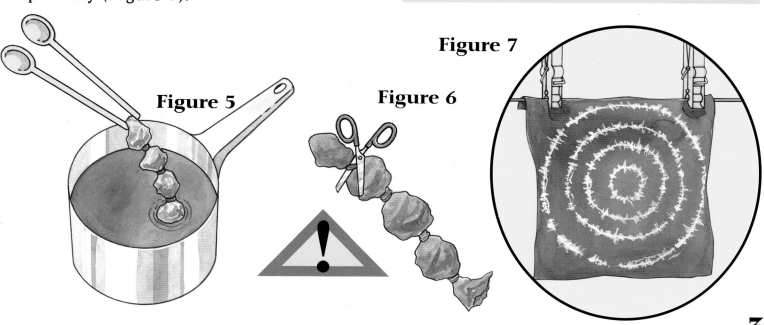

Figure 5

Figure 6

Figure 7

Colors and dyes

SPLITTING COLORS

Several different pigments are used to make the inks in colored pens. Cut circles out of paper towels (Figure 1). Draw a blob of color in the middle of each one with non-waterproof felt pens (Figure 1).

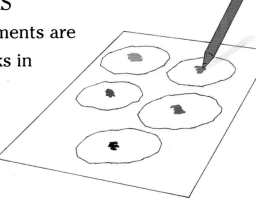

Figure 1

Put a drinking straw in water and cover the top with your finger (Figure 2). Remove and take your finger off to release a drop of water onto each blob of color (Figure 3). Watch the color spread out and you will see new colors appear (Figure 4).

Figure 2

WHAT THIS SHOWS

The water travels through the absorbent paper towels and spreads the pigments in the ink. It spreads each pigment at a slightly different speed. This lets you see all the separate pigments that make up each colored ink.

Figure 3

Figure 4

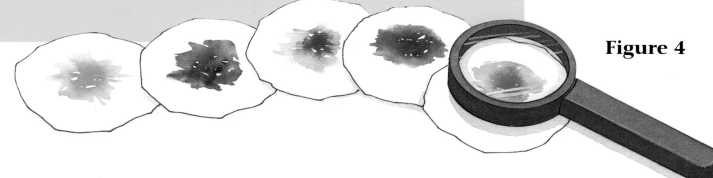

Beetle juice

Cochineal is a bright-red dye made from grinding small, red beetles. It is used as a food dye, so you have probably eaten it.

THE AMAZING MAGIC DOTS
Use dots to create a whole picture

Draw a picture just using dots of color. Use blue and yellow dots together where you want green; yellow and red dots for orange; and blue and red for purple. Stand back and see the colors and shapes appear.

WHY IT WORKS

Your brain mixes the colored dots for you, and you see the new colors.

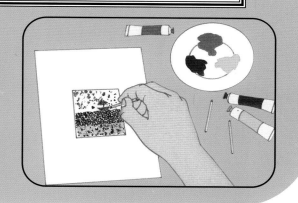

COLOR WHEEL

Here is how to make a color wheel. Paint in the primary colors—red, yellow, and blue. Mix them to make the secondary colors—orange, green, and purple. When you put colors opposite on the wheel next to each other, they can look much brighter.

Colors reflected from pigments are different from the colors of the spectrum. We see colors when pigments reflect certain colors of light into our eyes.

Taking pictures

Every day, we see still and moving images from all over the world in newspapers and on television. These images are captured by cameras on film and video, and with digital technology. We have family photograph albums and home movies. X-ray cameras take pictures of bones inside our bodies, and infrared cameras take pictures of light we can't normally see. Our eyes work like a small and very precise camera, that lets us see the world around us.

Make a pinhole camera

METHOD NOTES
The brighter the object you choose, the better the image will be in your camera.

Materials
- transparent tape
- tracing paper
- a small square box
- a pin
- a dark towel
- a magnifying glass

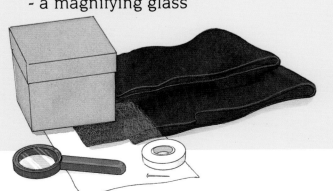

Figure 1

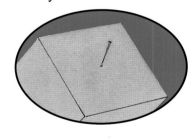

Figure 2

1. Tape the tracing paper smoothly over the open side of the box (Figure 1).
2. Make a pinhole in the middle of the side of the box opposite the tracing paper (Figure 2).

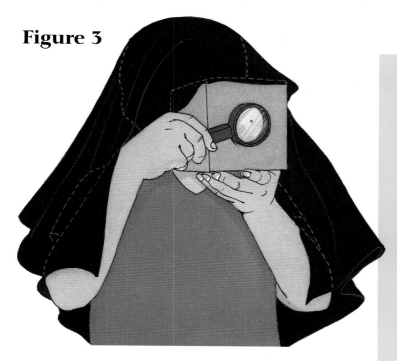

Figure 3

WHY IT WORKS

Light rays travel in straight lines. Rays from the top and bottom of the object cross over as they pass through the pinhole, and reverse the image onto the tracing paper, so you see it upside-down. In a camera, the image is projected onto photographic film, which contains chemicals that preserve it.

3. Put the towel over your head and the box to block out the light (Figure 3).
4. Point the pinhole at something very bright and look at the tracing paper.
5. Hold a magnifying glass between the pinhole and the object you are looking at for a clearer, bigger image.
6. You will see an upside-down image of the bright object reflected onto the tracing paper (Figure 4).

Your eyes work in the same way. A lens at the front of your eye focuses light onto light-sensitive cells in the retina at the back of your eye. An upside-down image is sent to the brain—which turns it around again.

Figure 4

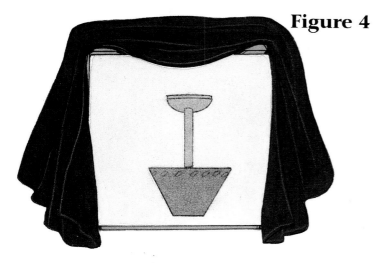

Taking pictures

Our eyes are able to see only a small range of light, called visible light. Cameras can be designed to capture light that is invisible to the human eye.

FIND OUT HOW X RAYS WORK

X rays can travel through soft things like skin or clothes. They are used to check inside luggage at airports and in hospitals to see broken bones. Draw around your hand on some white cardboard and cut the shape out (Figure 1). Stick the silhouette of your hand between two sheets of white paper (Figure 2). Look at the paper—you can't see the hand (Figure 3). Hold the articles up and shine a bright light behind them. You can see the outline of the hand clearly now (Figure 4).

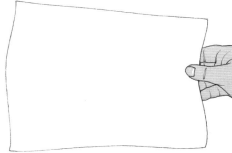

Figure 1

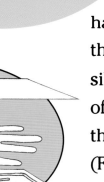

Figure 2

Figure 3

Figure 4

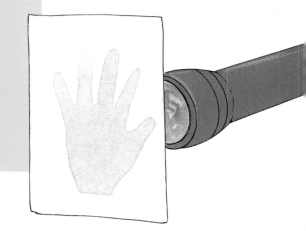

WHY IT WORKS

Visible light is a small part of the electromagnetic spectrum. Gamma rays, X rays, ultraviolet rays, visible-light rays, infrared rays, microwaves, and radio waves are all electromagnetic waves with different wavelengths. X rays have a much shorter wavelength than visible-light rays. X-ray light goes through our skin but not our bones, and we see shadows of our bones.

THE AMAZING FISH IN A BOWL
See how light can trick your eyes

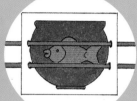

You need a piece of thick cardboard 2½ x 2½ in with two holes on opposite edges. Loop 2 ft of string through the holes. Draw a fish on one side of the cardboard and a bowl on the other. Twist the ends of the string, then pull each end of the string so the cardboard spins rapidly.

WHY IT WORKS

The fish appears to be in the bowl because the two pictures move so quickly that your eyes see them as one picture.

Hot cat

Infrared cameras capture invisible light called infrared radiation, which is given out by anything hot. They can take pictures of things that would usually be invisible at night.

MAKE A FLIP BOOK

Cartoons are made from sequences of pictures moving so quickly that our brain sees them as a moving image. Draw a sequence of pictures of a stick figure running. Staple the pictures together and flip the pages rapidly. You will see your figure running.

Cameras
focus light through lenses to capture images on film in the same way that lenses in our eyes focus light onto light-sensitive cells at the back of our eyes.

Weird light

The Sun, lamps, flashlights, and candles are familiar luminous objects that give out light. Light also comes from other, stranger sources. You can see flashes of light caused by static electricity when you take off a nylon shirt; fireflies flash out light signals at night; fluorescent strips shine in the glare of headlights; and chemicals glow inside light sticks. Lasers control and use the power of light, and light travels along fiber-optic cables carrying information.

Make light bend and pour to see how fiber–optics work

METHOD NOTES
Make the pinhole and the clear patch about one quarter of the way up from the bottom of the bottle.

Materials
- scissors
- a clear plastic bottle
- black paint
- a flashlight
- a pin
- a large glass bowl

Figure 1

1. Cut the top off the plastic bottle (Figure 1).
2. Paint the outside black, leaving a small, clear patch on one side (Figure 2).
3. Make a small hole in the bottle with the pin on the side opposite the clear patch (Figure 2).

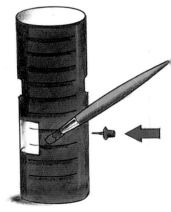

Figure 2

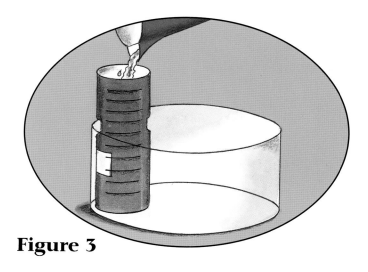

Figure 3

4. In a dark room, stand the bottle at one side of the bowl, with the clear patch facing out and the hole facing the bowl (Figure 3).

5. Fill the bottle with water (Figure 3). Then, shine a light at the clear patch.

6. Pressure will push the water through the hole in a thin stream that will glow brightly with light from the flashlight (Figure 4).

Figure 4

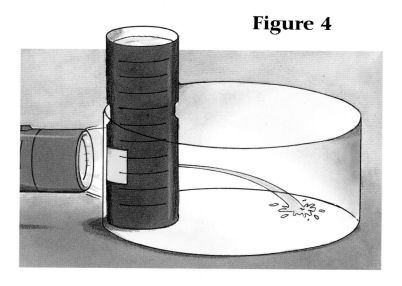

WHAT THIS SHOWS

Light carries information, such as telephone and television signals, along glass wires called fiber-optic cables. The cables channel light fast and efficiently around bends and for long distances. The stream of water acts like a fiber-optic wire. Light rays bounce off the sides of the stream and reflect inside it, making it glow.

Fiber-optic cables carry light along their path by reflecting it back and forth off the sides of the thin wires.

43

Weird light

MAKE LIGHT WITH SUGAR CUBES

Grains of sugar are tiny crystals. Sugar cubes are made by pressing grains of sugar together. Put some sugar cubes in a clear plastic food bag and tie the top (Figure 1).

Put the bag on a chopping board and darken the room. Smash the bag with a wooden rolling pin, and you will see sparks of light coming from the crystals of sugar (Figure 2).

Figure 1

Figure 2

WHY IT WORKS

Light is produced when crystals are rubbed together. Hitting the sugar with the rolling pin rubs the grains of sugar together with enough force to produce sparks of light.

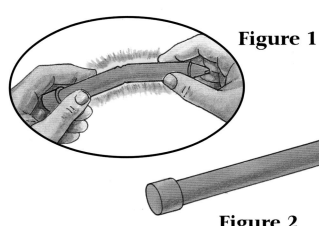

Figure 1

Figure 2

DISCOVER WHY PARTY STICKS GLOW

There are two different chemicals, one in an inner and one in an outer tube of a party glow stick. When you snap the tube (Figure 1), the chemicals mix and glow. You can stop a stick from glowing (Figure 2) by putting it in the freezer, and start it glowing again by dipping it in warm water.

Worms that glow

Glowworms are like party glow sticks glowing in the dark. Chemicals in their bodies mix to make a greenish light. They have the ability to turn their "lights" on or off any time they choose.

PAINT PICTURES THAT GLOW

Luminous paint works by storing light. When it becomes dim, you can recharge it by leaving it in bright light again. Make two pictures, one with luminous and one with ordinary paint or pens. Put the pictures in shade or dull light, and you will see how luminous paint glows with stored light.

Many different kinds of light illuminate the world. The Sun is the Earth's main source of light. Without light and heat from the Sun, there would be no life on Earth.

Glossary

concave lens

A curved piece of glass that spreads out light rays. It makes things appear smaller.

convex lens

A curved piece of glass that brings light rays together. It makes things appear larger.

fiber-optic cable

A very thin piece of glass along which light can travel long distances, carrying huge amounts of information such as telephone, television, and internet signals.

light ray

A narrow beam of light.

microscope

An arrangement of lenses and mirrors that can magnify things invisible to the naked eye.

opaque

The description of a material such as brick through which light cannot pass.

pigments

Substances that give paints and dyes their color.

primary colors

Colors which can be combined to make all other colors. The primary colors of light are red, green, and blue. The primary colors of paints are red, yellow, and blue.

prism

A transparent wedge of glass that refracts (bends) light, splitting it into all the colors of the rainbow.

reflection

The way light bounces off a surface. It reflects off a flat, smooth surface at the same angle it strikes it.

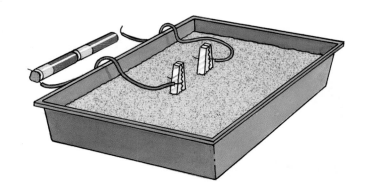

refraction

The way light bends when it passes from one material (such as air) to another (such as water).

shadow

A dark area formed on a surface that takes the shape of an object that has blocked light rays from the Sun or another light source.

spectrum

All the colors of visible light, from those with the shortest wavelength (violet) to those with the longest (red).

telescope

A scientific instrument that uses lenses and mirrors to magnify objects in the night sky, such as stars and galaxies.

translucent

The description of a material such as tracing paper that lets some light pass through it.

transparent

The description of a material such as glass that lets all or nearly all light pass through it.

wavelength

Light travels in waves. A wavelength is the distance from the top of one light wave to the top of the next. We see lights of different wavelengths as different colors.

X rays

A kind of wave that cannot be seen by the human eye. X rays can pass through some materials that are opaque to visible light.

Index